# JACKALS EAT ROTTING FOOD!

Natalie Humphrey

**Please visit our website, www.garethstevens.com. For a free color catalog of all our high-quality books, call toll free 1-800-542-2595 or fax 1-877-542-2596.**

**Library of Congress Cataloging-in-Publication Data**
Names: Humphrey, Natalie, author.
Title: Jackals eat rotting food! / Natalie Humphrey.
Description: Buffalo, New York : Gareth Stevens Publishing, [2024] | Series: Nature's grossest | Includes index.
Identifiers: LCCN 2022056730 (print) | LCCN 2022056731 (ebook) | ISBN 9781538285756 (library binding) | ISBN 9781538285749 (paperback) | ISBN 9781538285763 (ebook)
Subjects: LCSH: Jackals–Juvenile literature.
Classification: LCC QL737.C2 H857 2024 (print) | LCC QL737.C2 (ebook) | DDC 599.77/2–dc23/eng/20221129
LC record available at https://lccn.loc.gov/2022056730
LC ebook record available at https://lccn.loc.gov/2022056731

Published in 2024 by
**Gareth Stevens Publishing**
2544 Clinton Street
Buffalo, NY 14224



Designer: Leslie Taylor
Editor: Natalie Humphrey

Photo credits: Series art (background) Oleksii Natykach/Shutterstock.com; cover Henk Bogaard/Shutterstock.com; p. 5 Robert Henry Hurn/Shutterstock.com; p. 7 (top) Four Oaks/Shutterstock.com, (bottom left) Wim Hoek/Shutterstock.com, (bottom right) Chris Fourie/Shutterstock.com; p. 9 Martin Mecnarowski/Shutterstock.com; p. 11 Cathy Withers-Clarke/Shutterstock.com; p. 13 Tobie Oosthuizen/Shutterstock.com; p. 15 Tobie Oosthuizen/Shutterstock.com; p. 17 Stephen Barnes/Shutterstock.com; p. 19 Cathy Withers-Clarke/Shutterstock.com; p. 21 JMx Images/Shutterstock.com.



Printed in the United States of America

CPSIA compliance information: Batch #CSGS24: For further information contact Gareth Stevens at 1-800-542-2595.

# CONTENTS



**Boldface** words appear in the glossary.

## Cute Canines

Jackals are a cute **canine** that will eat anything! This small animal loves anything it can get its **jaws** on. This usually means fresh meat and berries. But jackals eat gross food too. Hungry jackals eat rotten meat!

## Kinds of Jackals

There are three **species** of jackals. Black-backed jackals and side-striped jackals are both found in Africa. The golden jackal, or common jackal, is found in Africa, south and Southeast Asia, and Europe. Jackals love anywhere they can hide!

BLACK-BACKED

## Small Dogs

Jackals aren't very big. They're usually 16 inches (40 cm) tall and up to 33 inches (84 cm) long. They weigh between 11 and 26 pounds (5 to 12 kg). That's around the size of a Shetland sheepdog.

## Hiding Their Smell

Jackals can be pretty gross. When they're hunting, jackals will roll on the ground to hide their smell. This means jackals will roll in dirt, grasses, or even poop. Their **prey** can't smell them coming until it's too late!

## Pack Hunters

Jackals live with their families, or packs. Packs are made of two parents and their babies. They may hunt together. When jackals work together, they can take down bigger animals.

## Hunting Alone

Jackals hunting alone can't catch big meals. Instead, one jackal might eat small lizards, **rodents**, birds, or bugs. They will also eat plants such as grasses or berries. Jackals aren't picky and will eat nearly anything they can catch.

## Food Thieves

Jackals don't just hunt for their food. They also steal meals from bigger predators. When a hyena takes down a meal, hungry jackals will sneak up and try to grab a piece while the hyena is eating.

HYENA
JACKALS

## Gross Dinner

Jackals that can't find fresh food aren't picky. If they're hungry, they'll eat anything a predator has left behind. Jackals will even eat rotten food that's been left sitting in the hot sun for days!

## Special Stomachs

Rotten food doesn't just look gross. It can make some animals very sick. But **scavengers** that eat rotten meat have special **bacteria** in their stomach that protect them. These bacteria help keep scavengers, like jackals, safe from the food they eat!

# GLOSSARY

**bacteria:** Tiny creatures that can only be seen with a microscope.

**canine:** The name scientists use for the dog family.

**jaw:** One of the two bones of the face where teeth grow.

**prey:** An animal that is hunted by other animals for food.

**rodent:** A small, furry animal with large front teeth, such as a mouse or rat.

**scavenger:** An animal that eats the remains of dead animals.

**species:** A group of plants or animals that are all of the same kind.

# FOR MORE INFORMATION

## BOOKS

Humphrey, Natalie. *Jackals in the Wild.* Buffalo, NY: Gareth Stevens Publishing, 2023.

Stewart, Melissa. *Ick! Delightfully Disgusting Animal Dinners, Dwellings, and Defenses.* Washington, DC: National Geographic Kids, 2020.

## WEBSITES

**African Wildlife Foundation: Jackal**
*www.awf.org/wildlife-conservation/jackal*
Learn more about jackals and how to protect them in the wild!

**A to Z Animals: Jackal**
*www.a-z-animals.com/animals/jackal/*
Check out more interesting jackal facts.

# INDEX